Food
114

准备好冒险了吗?

Ready to Take a Risk?

Gunter Pauli

[比] 冈特·鲍利 著

[哥伦] 凯瑟琳娜·巴赫 绘

颜莹莹 译

上海远东出版社

丛书编委会

主　任：田成川

副主任：闫世东　林　玉

委　员：李原原　祝真旭　曾红鹰　靳增江　史国鹏

　　　　梁雅丽　孟小红　郑循如　陈　卫　任泽林

　　　　薛　梅　朱智翔　柳志清　冯　缨　齐晓江

　　　　朱习文　毕春萍　彭　勇

特别感谢以下热心人士对童书工作的支持：

匡志强　宋小华　解　东　厉　云　李　婧　庞英元

李　阳　梁婧婧　刘　丹　冯家宝　熊彩虹　罗淑怡

旷　婉　王靖雯　廖清州　王怡然　王　征　邵　杰

陈强林　陈　果　罗　佳　闫　艳　谢　露　张修博

陈梦竹　刘　灿　李　丹　郭　雯　戴　虹

目录

Contents

老鼠和鸽子想知道他们是否能得到充足的食物——不光是今天和明天的，甚至还有明年的。他们的家庭成员越来越多，一些家人开始担心粮食可能不够。

"你知道，如果我们老鼠家族的成员越来越多，粮食却没有增多，那么食物就会越来越贵。"老鼠分析道。

A rat and a pigeon want to find out if there is enough food for everyone – for today, tomorrow and for next year. Their families are growing larger and some are worried that there may not be sufficient food for everyone.

"You know, when there are more and more rats, and the amount of food stays the same, then the food will be more expensive," comments Rat.

是否每个人都能有充足的食物？

Is there enough food for everyone?

继续饿下去？

Go hungry for a couple of days?

"是啊，这样我们当中能买得起食物的就会越来越少。那我们该怎么办呢？继续饿下去？"

"哦，不，我可没有耐心一直干等，也不想眼睁睁看着孩子们挨饿却什么都不做。"

"你要知道，明年才有新的收获季呢。"鸽子说。

"True, and then less of us can afford it. So what is the option? Go hungry for a couple days?"

"Oh no, I do not have the patience to wait until there is more food, and do not want to see my children suffering, without me taking any action."

"You know, the new harvest will only be ready next year," says Pigeon.

"那不就意味着我们要一直饿到明年？我们家刚出生的小宝宝们可撑不到那时候！"

"也许你在期待明年会有更多食物。要是明年食物没那么多呢？"

"那我们就只好搬去别处了，去一个有食物的地方。"老鼠回答道。

"Does that mean we will have to go hungry until next year? None of our newborn little ones will survive!"

"You are hoping there may be more food next year. What if we know that there is not going to be more food?"

"Then we will simply have to move to another place, one where there is food," Rat replies.

我们家刚出生的小宝宝们一个都不能……

None of our newborn little ones ...

我不喜欢孤注一掷……

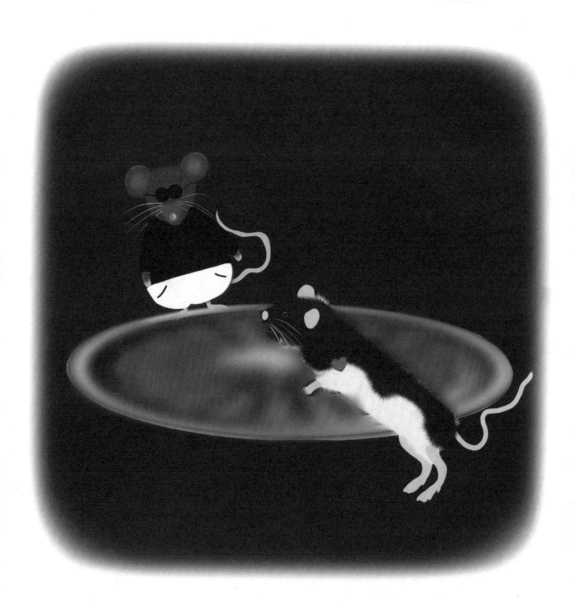

I hate to gamble ...

鸽子又向老鼠提问："为了确保你能有充足的食物，你会把所有的钱都用来买食物吗？"

"我不喜欢孤注一掷。"老鼠说，"我知道如果我花掉所有积蓄，那么接下去几个月就有吃的，但我还不知道明年会怎样。"

"那你会怎么选择？"

Pigeon then asks him another question, "Are you ready to use all your savings for food, so that you can be sure that you have enough?"

"I hate to gamble," Rat says. "I know that if I use all my savings, I can get food for the next few months, but I will still have no idea what next year will bring."

"What are your options?"

"我可能会收拾一下然后搬走。"老鼠说，"去一个大家都能有食物的地方。"

"要我说，你这样做也是一个大冒险。"

"可是你要知道，我们不论做什么都是在冒险。就算我们什么都不做，我们也没法保证什么。"

"I could pack my bags and move away," Rat says. "Go somewhere where there is food for all."

"If you ask me, that is a big risk as well."

"You know, we take risks whatever we do. Even if we do nothing, we can never be sure."

我可能会收拾一下然后搬走……

I could pack my bags and move away ...

谁又知道他们有没有足够多的种子？

Who knows if they have enough seeds?

"那些卖给我们食物的其实也在冒险。"鸽子说。

"怎么会呢？他们今天卖出去，就能拿到钱，然后就能做他们想做的事情。"

"是的，但是当他们知道明年我们的数量会增多，他们今年就会种植更多粮食，明年就会赚更多钱。"

"谁说他们有土地种植更多粮食？谁又知道他们有没有足够多的种子？"

"And those who sell the food to us are also taking risks," says Pigeon.

"Why? They sell today, get their money, and can do with it what they want."

"Yes, but when they know that there are more of us next year, then they can plant more now, and earn more money next year."

"Who says they have the land to plant more? Who knows if they have enough seeds?"

"呃，但愿他们有吧。"鸽子说，"如果他们既没有地也没有种子，我们就都麻烦了。"

"也许他们会留一些存粮，以备不时之需。"

"嗯，我相信他们有。如果他们把存粮卖了，接下去几个月我们就有吃的了。但是如果他们不马上卖，而是等得久一些，也许我们会因为太饿而愿意付高价给他们，这样他们用同样多的粮食便可以赚取更多的利润！"

"Well, let's hope that they do," Pigeon says. "Because if they do not have the land nor the seeds, then we are all in trouble."

"But maybe they have some extra food kept aside, just in case."

"Well, I am sure they have. If they sell that, then we'll have food for the next few months. However, if they wait a little longer, then we may be so hungry that we will be prepared to pay a lot more, and then they make much more profit from the same amount of food!"

......用同样多的粮食便可以赚取更多的利润!

... more profit from the same amount of food!

这太不道德了！

That is just not ethical!

"你的意思是他们囤积粮食不急着卖，当我们越来越迫切地需要粮食时他们便高价卖出？"老鼠问道，"这太不道德了！"

"是啊，可是如果你决定了要举家搬迁，就算忍饥挨饿也带着你全部的积蓄离开，那他们就要守着那些也许永远都卖不出去的存粮了。这样他们就输了。"

"唉，这可真糟糕。"

"You mean keeping food for later, and selling it at higher prices when we are more desperate?" asks Rat. "That is just not ethical!"

"Yes, but if you decide to leave with your whole family and take your money and your hunger with you, then they are stuck with a reserve they may never be able to sell. Then they lose."

"Gosh, this is a difficult situation."

"我了解。而且虽然我并不想冒任何风险，但是作为一只鸽子这对我来说很容易，我可以飞得远远的。毕竟如果你感到饿，你有权利变得失去耐心，"他继续道，"但如果这会危及你孩子们的健康和幸福的话，还是不要冒险了吧。"

"你说得很对。"老鼠答道，"那么，如果你确定要搬走，记得寄明信片给我。如果你找到食物充足的地方，一定要告诉我，我们全家去找你。"

"那太棒了！"

……这仅仅是开始！……

"I know, and while I do not want to take any risks either, as a pigeon it is easy for me, I can fly far away. One has the right to be impatient if you are hungry," he continues, "but when your kids' health and happiness are at stake, never take a risk."

"You are right about that," Rat replies. "Well, send me a postcard if you do move away. When you find land with more food, let me know and we will all come and join you."

"That will be great!"

... AND IT HAS ONLY JUST BEGUN! ...

·····这仅仅是开始！······

... AND IT HAS ONLY JUST BEGUN! ...

Did You Know?

你知道吗？

Speculation implies a risk of loss, which is expected to be more than offset by the possibility of a big gain.

投机意味着承担损失的风险，人们往往期待高收益会抵消这个风险而有盈余。

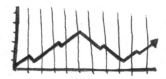 

Speculating is to profit from the direction (up or down) the price or value of an asset or a commodity moves, assuming risk. Hedging aims to reduce risk by taking an opposite position than the market is expected to follow, cancelling our profits or losses.

投机是利用资产或商品的价格或价值浮动而从中获益，并承担风险。对冲的目的是站在与市场预期相反的角度降低投机的风险，从而使收益或损失都减少。

套利是利用市场效率低下导致的不同股票或外汇交易的不匹配，同时买进卖出，以期在价格差中获利的行为。

Arbitrage is the simultaneous buying and selling in order to profit from the small differences in price, exploiting the mismatch in different stock or currency exchanges due to market inefficiencies.

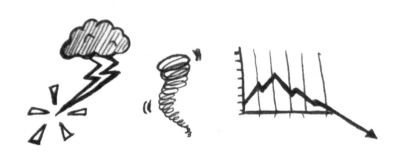

风险分析是指对由潜在的自然或人为造成的不利结果的危险性进行定义。

Analysing risk is defining the dangers posed by potential natural and human-caused adverse events.

We are good at weighing risks and benefits of doing something new, but often are unwilling to analyse the risk of doing nothing at all.

我们总是善于权衡尝试新事物的利弊，却往往不愿意分析什么也不做会带来多大的风险。

Rational self-interest is an economic principle that describes bahaviours that promote one's own interests through economic decisions. However, decisions are often guided by perceived fairness and loss aversion, which are emotions and attitudes, and not logic.

理性利己主义是一种经济原则，指通过经济决策利好自身。然而人们的决策往往由感知公平、规避损失这些非逻辑性的情感和态度所引导。

An option gives the buyer a right – but not an obligation – to buy or sell an asset at a specific price. A futures contract gives the buyer the obligation to purchase an asset, and the seller to sell at a specific future date.

期权使买方享有能够以一定金额买卖某项资产的权利，而非义务。期货合约使买方有义务购买某项资产，也使卖方有义务在未来某一特定时间卖出。

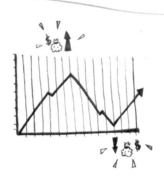

Speculating on a bullish market means that you expect prices to go up, speculating on a bearish market means that you expect prices to go down. But, money could be made from both: from prices going up or prices coming down.

投资牛市表示你期待价格上涨，投资熊市表示你期待价格下跌。不管价格涨或跌，都有可能从中赚钱。

Do you think it is possible to have all information with full certainty before you make a decision?

你认为有没有可能在你做决定之前确保掌握所有的信息？

Would you be ready to change for something with an uncertain outcome, when you are uncertain about what you have now?

当你对你的现状还没有把握时，你会为了某件不确定结果的事情而做出改变吗？

If there was a food shortage, would you hold on to your savings in cash, or would you prefer to use your savings to get a secure food supply?

假设出现食物短缺，你会坚持继续持有现金，还是会将现金换成安全的食物供给？

What is the difference between taking a risk and gambling?

冒险和赌博有什么不同？

Do It Yourself!

自己动手!

You are stranded on an island and your call for help is answered. The helicopter that is sent to rescue you, may not have enough fuel to get back; the boat that came to your rescue has hit the coral and is taking on water. What would you do? Remain on the island, board the helicopter or swim to the boat? Discuss the options you have and the risks that you are taking, by doing something, or doing nothing at all.

假设你被困在一座岛上，你发出求救信号并得到回应。派去救援的直升机有可能会因燃料不够而无法返回；前去营救你的船也因触礁而进水。你会怎么办呢？是继续待在岛上、搭乘直升机还是游到船上去？讨论一下，行动或是不行动，不同的选择会有哪些不同的风险？

学科知识
Academic Knowledge

生物学	根据四季的气候变化,收成是周期性的,但在热带地区是持续性的。
化 学	只要控制好温度、压力和催化剂等反应条件,化学反应的风险是很低的。
物 理	由于物理定律没有例外,物理定律在执行时会完全按照预期。
工程学	对基础设施、信息系统、制造业或农业的任何一项投资都需要进行风险分析,验证什么可能出错。
经济学	投机和套利的区别;对冲和投机的区别;从心理研究到经济科学的洞察表明,不确定情况下的人为判断和决策不总是理性的;信息不对称,交易的一方拥有比另一方更优化的信息;为确保未来收益而进行再投资的需求;储蓄和投资之间的平衡;在做出理性决策时假设的重要性;供需平衡;预期的重要性:我们预期会发生什么,就可能会影响它的发生;市场预测:研究未来需求在哪里,以便相应地调整生产和投资;货物、人和金钱流动的影响。
伦理学	不把食物卖给当下正需要它的人,而是等到粮食短缺、饥饿人数增加的时候再卖,以此赚取更多的钱,这样做十分不道德!
历 史	心理学家丹尼尔·卡尼曼因指出在不确定性下的决策是由情感而非单由理性所引导,获得2002年诺贝尔经济学奖。
地 理	经济学运用国家风险这一概念,指与一个国家发生贸易、投资或借贷关系所产生的风险,这些风险可能会因这个国家政治、汇率和债务偿还能力的变化而显现。
数 学	在定量风险分析中,尝试用数值确定概率;定性方法包括定义各种威胁、确定漏洞的程度并在出现问题时制定对策;博弈论是对智慧、理性的决策者间应对冲突与合作的数学模型的研究;非合作博弈论;概率论涉及随机现象的分析;模拟真实或接近真实的生活事件以找到过去发生问题的原因(例如事故),或对假设的情况或因素下将会产生的影响或结果做出预测;模拟需要数学模型,用来构建物理模型、阶段演练或计算机图形模型(诸如动画流程图)。
生活方式	为应对困难时期而储蓄;赌博的欲望。
社会学	食物短缺或就业机会减少导致人口流动。
心理学	与年轻人比,年长者面对潜在得失更倾向于规避风险;抵抗不确定因素导致的压力的能力;我们认为从事新事物会承担风险,殊不知无所事事或重复同样的事情也有风险,特别是当外部环境变化超出我们控制时。
系统论	现实世界中,决策制定通常在不完全获取信息的情况下进行,这就迫使我们做出假设及猜想,跟随情感引导。这使得决策复杂化,但现实就是如此,因为我们生活在一个充满不确定性的时代。

情感智慧
Emotional Intelligence

老鼠

　　老鼠以一种理性的方式开始对话，逻辑清晰，但后来有点失去耐心。他无法一直等待，除非确定有更多的食物可供日渐增多的家庭成员生存下去。他在评估等待的风险，这可能导致一个戏剧性的结果：他的孩子们可能会没有食物。老鼠还分析了其他选择，他想到了搬走，但前提是他得确定那个地方有足够的食物。然而这一点也是非常不确定的。他意识到做事情必有风险，也总结出什么都不做也是一种风险。老鼠站在食物卖方的立场，从另一个角度看到事实，那就是他无法获取全部的信息。老鼠认为如果有人挨饿，就必须卖出食物，那些囤积粮食以赚取高价的投机行为是不道德的。

鸽　子

　　鸽子缺乏决断力，尽管他能够洞察一些常识，也会问很多问题。鸽子遵循老鼠的逻辑，帮其梳理各种设想，显示出他的同理心。鸽子促使老鼠从不同的角度看待现实。他提出不同视角，并指出老鼠的决定会带来哪些影响。鸽子具有自知之明，他意识到，与老鼠相比，他能够很容易地做出决定，因为他可以轻易地飞走，而老鼠却要举家搬迁，走向一个不确定的未来。

艺术
The Arts

　　艺术家就像社会的触角。在那些充满未知的年代，艺术家用作画的方式表达他们眼中的理性世界，这些不同的印象表达基于他们对信息的收集或猜想。你会怎样描绘未知呢？是画一个乌云层叠的灰色天空？还是能想出其他的方式来表现这种迷惘感——有很多选择但不清楚未来会把我们带向何处，或者有很多条路，但却不知道选择哪一条？

思维拓展
Systems: Making the Connections

生活中唯一确定的就是改变。人类喜欢建立一个充满确定性的生活，然而实际上我们对任何事情都没有把握。万物都是变化着的，几乎没有预兆。此外，人类带来很多改变，可是我们却不愿意改变自身行为来适应新环境，哪怕这多少是由我们自己创造的。持续的改变使我们必须适应并做出选择，在多个选项中做出决定。也许我们可以选择某条路，有明确的目标，但是却不能保证结果会像我们所期待的那样万无一失。风险是生活的一部分，我们可以努力降低风险，却不能消除它们。情景模拟是了解我们的备选方案的一个办法，也就是问自己："如果……会怎样呢？"持续处于不同的情境中可以使我们熟悉将来可能会面临的各种决定。当我们能够持不同的观点看待同一个问题的时候，情境会更加显见易懂，形成数学模型或演练（就像军队的军事演习一样）。如果我们待在老地方，而家庭成员变多，我们就需要更多的食物供给。如果我们搬去别处定居，那些食物供应商就会面临需求的减少，可能会剩下许多食物。商业和金融的世界不容许赌博，因而衍生出了一些技巧用于理解这些复杂的互动关系，也创造出减少风险的工具，包括对冲。家庭单位不会使用这些复杂的工具，而是更倾向于储蓄，这样万一经历困难时期，足够的积蓄可以缓解对家庭造成的影响。

动手能力
Capacity to Implement

把你当下正面临的一个挑战写下来，然后写下你为了减少或降低风险而做的决定。用思维导图呈现出所有可能的选项。一旦你能够用图表的方式清晰地列出所有不同的选择以减缓、规避或克服挑战及相关风险，就把你的决定分享出来。运用你的思维导图，让他人理解你的逻辑和决策过程。要记住你的许多选项并非线性的，可以任意更改，它们都是你对现实世界感知的反映。

故事灵感来自
This Fable Is Inspired by

约翰·纳什
John Nash

约翰·纳什于 1928 年出生在美国西弗吉尼亚州。在他年轻的时候，他的父母鼓励他学习高等数学。他 19 岁毕业于卡内基理工学院，获得数学硕士学位。纳什凭其博弈论论文获博士学位，博弈论是用数学的方法对智慧、理性的决策者之间冲突与合作模型的研究。纳什探讨了在日常生活里常见的复杂系统中对变化与决策起决定作用的因素。他的理论被广泛应用于经济学。由罗素·克劳主演的奥斯卡获奖影片《美丽心灵》就是以有关他的生活、工作及健康状况的故事为原型改编的。

图书在版编目（CIP）数据

冈特生态童书.第四辑:修订版:全36册:汉英对照 /
（比）冈特·鲍利著;（哥伦）凯瑟琳娜·巴赫绘;
何家振等译.—上海:上海远东出版社,2023
书名原文: Gunter's Fables
ISBN 978-7-5476-1931-5

Ⅰ.①冈… Ⅱ.①冈… ②凯… ③何… Ⅲ ①生态环
境–环境保护–儿童读物—汉、英 Ⅳ.①X171.1-49

中国国家版本馆CIP数据核字(2023)第120983号
著作权合同登记号图字09-2023-0612号

策　　划	张　蓉
责任编辑	张君钦
封面设计	魏　来　李　廉

冈特生态童书

准备好冒险了吗？

[比]冈特·鲍利　著
[哥伦]凯瑟琳娜·巴赫　绘

颜莹莹　译

记得要和身边的小朋友分享环保知识哦！
八喜冰淇淋祝你成为环保小使者！